INTERNATIONAL CENTRE FOR MECHANICAL SCIENCES

COURSES AND LECTURES No. 71

AF065608

D.R. AXELRAD
McGILL UNIVERSITY MONTREAL

RANDOM THEORY OF DEFORMATION OF STRUCTURED MEDIA

D.R. AXELRAD - J.W. PROVAN

THERMODYNAMICS OF DEFORMATION IN STRUCTURED MEDIA

COURSE HELD AT THE DEPARTMENT
OF MECHANICS OF SOLIDS
JULY 1971

UDINE 1971

SPRINGER-VERLAG WIEN GMBH

Originally published by Springer - Verlag Wien - New York in 1972

ISBN 978-3-211-81175-7 ISBN 978-3-7091-2936-4 (eBook)

DOI 10.1007/978-3-7091-2936-4

D. R. AXELRAD

RANDOM THEORY OF DEFORMATION OF STRUCTURED MEDIA

P R E F A C E

A theory of deformation of structured media derived from statistical mechanics and the theory of probability is introduced. The basic deformation kinematics of a simple model and a "material functional" that replaces the conventional constitutive relations are discussed. Functional analysis is used for the description of the field equations in the function space.

Udine, July 1971

D.R. Axelrad

1. Introduction

In the last decade several theories have been proposed with the aim of modifying or extending classical continuum theory so that the deformation process of structured media can be described. The first theory considering the presence of a microstructure known as the theory of 'oriented media' is due to E. and F. Cosserat [1]. In this theory the deformation is described in terms of a position vector of an arbitrary point in the medium with respect to a fixed reference frame and a vector called "director" associated with the position vector. The concept of using directors in continuum mechanics goes back to Duhem [2]. The fundamental aspects of the deformation kinematics of such media were treated comprehensively by Truesdell and Toupin [3]. Following the Cosserat approach, Mindlin [4] proposed a theory of elastic media possessing a microstructure in which a physical point or "unit cell" was considered deformable. This theory can be reduced in the case of a homogeneous deformation to a model suggested by Ericksen and Truesdell [5], which is based on the concept of a Cosserat continuum. Another theory extending the classical formulation is the "couple stress theory" treated in detail by Toupin [6]. This approach was later completed by Eringen and Suhubi [7]. These researches also introduced a more general theory of microcontinuum that has been extended more recently to the continuum theory of "micromorphic media" Finally, Green and Rivlin [8] introduced the "multipolar" continuum theory, which uses force and stress multipoles that are defined in terms of the velocity components and their spatial derivatives. The above theories mentioned briefly here, represent the major contributions towards a modification of classical continuum theory.

However, it is the aim of this paper to discuss the "random theory of deformation of structured media" already proposed in several previous publications [9,10,11,12] . In this theory an ensemble of discrete particles is considered and the deformations are derived from concepts of statistical mechanics and the theory of probability. Hence, in order to characterize the micromorphology of the medium the relevant quantities are considered as random variables. It has been shown in the previous publications that in order to characterize the material response on a "local" level it is necessary to introduce the concept of a 'mesoscopic region", i.e., an intermediate domain with -in the material sample. By using this concept the transition from the local or microscopic description to the macroscopic one can be achieved. This concept is illustrated, in some examples, by Fig. 1. Another essential step in the analysis of the deformation of structured media is the notion of the existence of a "material functional", which contains the deformational and geometrical properties of the medium as functions of time and in general will also be temperature dependent. In a very reduced and simplified form such a functional can be written as follows:

$$\underset{\sim}{\mathfrak{F}}^{\tau} = \underset{\sim}{\mathfrak{F}}\left\{ A^{\tau}, T^{\tau}, \varphi^{\tau}, \ldots \right\}, \tag{1.1}$$

where A^{τ} is an operator representing material properties, T^{τ} the temperature level at time τ at which the the external force field is applied and φ^{τ} is associated with other parameters such as deformation gradients, orientation of the medium, etc., that are related to the microstructure and its changes during the deformation. For instance, if the parameter φ^{τ} is associated with the relative separation of two adjacent microelements α_1, α_2 within the mesoscopic domain, then from a purely geometrical point of view, it can be written as:

$$\varphi^{\alpha_1,\alpha_2} = \left| \underset{\sim}{r}^{\alpha_1} - \underset{\sim}{r}^{\alpha_2} \right|, \tag{1.2}$$

where $\underset{\sim}{r}^{\alpha_1}, \underset{\sim}{r}^{\alpha_2}$ denote the current positions of the center of mass of these two elements relative to a fixed reference frame. However, before discussing the deformation analysis on the basis of statistical mechanics and probability theory it is necessary to consider the kinematics of deformations first.

2. Kinematics and Microdeformations

Due to the multitude of observable microstructures, certain restrictions on the proposed random theory of deformation are required. Thus, physical parameters like dislocation density (metals) of the microelements, electrostatic potentials (soils) between them, heat flux, etc., as well as coupling effects are not included in the present formulation. In this context, reference is made to investigations at present being carried out concerned with the effect of dislocations on the response behaviour of polycrystalline solids [13] and the presence of surface potentials at the interface of crystals. For the study of the kinematics of a discrete particle ensemble, it is assumed that the microstructure of the medium can be represented by elements (single grains) of volume $\overset{\alpha}{v}$, boundary surface $\overset{\alpha}{s}$ on which a random microstress $\overset{\alpha}{\underset{\sim}{\xi}}$ is acting. In the subsequent analysis the six independent components of the microstress and microstrains (monopolar analysis) will be written in direct notation by noting that:

$$
(2.1) \qquad \begin{cases} \overset{\alpha}{\underset{\sim}{\xi}} = \overset{\alpha}{\xi}_{(ij)} , & \alpha = 1, \ldots, N , \\ \underset{\sim}{\epsilon} = \overset{\alpha}{\epsilon}_{(ij)} , & i, j = 1,2,3 . \end{cases}
$$

Their average values over the "mesoscopic domain" are similarly denoted by:

$$\underset{\sim}{\sigma} = \sigma_{(ij)} = \langle \underset{\sim}{\xi} \rangle_{V^N} \; ; \quad \underset{\sim}{e} = e_{(ij)} = \langle \underset{\sim}{\epsilon} \rangle_{V^N} . \qquad (2.2)$$

By introducing the concept of a mesoscopic subdomain of the macroscopic material body, where $\overset{\alpha}{v} \in \overset{N}{V} ; d = 1,\dots, N$ (N very large) it is assumed that the principles of statistical mechanics within this domain hold and that the macroscopic volume V of the medium is such that $V \gg \overset{N}{V} \gg \overset{\alpha}{v}$. This model can of course only be considered as a first approximation to an actual material structure. Furthermore, due to the inherent geometric inhomogeneities, "local" perturbations of the microstresses and strains are to be expected. Hence, locally one can write for $\overset{d}{\underset{\sim}{\xi}}$ and $\overset{d}{\underset{\sim}{\epsilon}}$:

$$\left.\begin{aligned} \overset{\alpha}{\underset{\sim}{\xi}} &= \underset{\sim}{\sigma} + \overset{*\alpha}{\underset{\sim}{\xi}} , \\ \overset{\alpha}{\underset{\sim}{\epsilon}} &= \underset{\sim}{e} + \overset{*\alpha}{\underset{\sim}{\epsilon}} , \end{aligned}\right\} \qquad (2.3)$$

where the starred quantities represent randomly varying stress and strain of zero mean value. These fluctuating components are a measure of the local deviation from the mean value and are significant in the study of yield phenomena of structured media[12] So far as the average values are concerned, if it is considered that the mesoscopic volume $\overset{N}{V}$ is intersected by a plane so as to disclose a total area $\overset{N}{A}$, then this area is also the sum of the intersected areas of certain of the microvolumes within $\overset{N}{V}$, say:

$$A^{N} = \sum_{\alpha=1}^{N} \overset{\alpha}{a} , \tag{2.3}$$

and by summation of the force resultants one obtains:

$$\underset{\sim}{\sigma} = \frac{1}{A^{N}} \sum_{\alpha=1}^{N} \overset{\alpha}{\underset{\sim}{\xi}} \overset{\alpha}{a} . \tag{2.4}$$

In view of the earlier assumption that a large number of areas $\overset{\alpha}{a}$ exist in a cross-section A^{N}, the components of the microstress tensor can be represented by the first moment of the distribution function of the random variable $\overset{\alpha}{\underset{\sim}{\xi}}$ and similarly for other quantities. A complete statistical description of the random quantities, however, would require the knowledge of all distribution functions with respect to any α-point system in the space of configurations. This difficulty may be overcome by using first and second moments of the relevant quantities only, which is the case in correlation theory [14].

Denoting the material position vector to the center of mass of microvolume $\overset{\alpha}{v}$ by $\overset{\alpha}{\underset{\sim}{R}}$ and the spatial position vector by $\overset{\alpha}{\underset{\sim}{r}}$, where both are taken with respect to the same Eulerian reference frames (see Fig. 2), then in general the motion of microelements can be represented by:

$$\overset{\alpha}{\underset{\sim}{r}} = \underset{\sim}{r} \left(\overset{\alpha}{\underset{\sim}{R}} , t \right) , \tag{2.5}$$

where $\underset{\sim}{r}^{\alpha}$ is a random variable. However, the precise motion of a material point of microelement is not known and therefore the representation as given in equation (2.3), will be adopted here also so that:

$$\underset{\sim}{r}^{\alpha} = \langle \underset{\sim}{r}^{\alpha} \rangle + \overset{*}{\underset{\sim}{r}}{}^{\alpha} , \tag{2.6}$$

where $\langle \underset{\sim}{r}^{\alpha} \rangle$ depends on the change due to the applied external force field and $\overset{*}{\underset{\sim}{r}}{}^{\alpha}$ are the fluctuations from this position. The latter occur in a random manner so that $\langle \overset{*}{r}{}^{\alpha} \rangle_{V^N} \equiv 0$. However, such a condition will only hold for the case of long range interactions or weak coupling between microelements. It is also considered here that $\langle \underset{\sim}{r}^{\alpha} \rangle \gg \overset{*}{\underset{\sim}{r}}{}^{\alpha}$ and that $\overset{\circ}{\underset{\sim}{R}}{}^{\alpha} \approx \langle \underset{\sim}{R}^{\alpha} \rangle$, where the latter quantity indicates a fixed value of $\underset{\sim}{R}^{\alpha}$ within the mesoscopic domain on the assumption that no significant variation of $\underset{\sim}{R}^{\alpha}$ in this domain occurs. This means that from a macroscopic point of view related to experimental observations, the mesoscopic region is rather small, but large enough to contain a large number of microelements.

The most important quantity in the kinematics of deformation is the microdeformation, which can be represented here as follows:

$$\left.\begin{aligned} \underset{\sim}{u}^{\alpha} &= \underset{\sim}{r}^{\alpha} - \underset{\sim}{R}^{\alpha} , \\ \underset{\sim}{u}^{\alpha} &= \langle \underset{\sim}{u}^{\alpha} \rangle + \overset{*}{\underset{\sim}{u}}{}^{\alpha} ; \quad \langle \overset{*}{u}{}^{\alpha} \rangle_{V^N} \equiv 0. \end{aligned}\right\} \tag{2.7}$$

Here the quantity $\langle \underset{\sim}{u}^{\alpha} \rangle$ and in particular its value on the boundary of the mesoscopic region is the microdeformation which, taken as its average value will be the microscopic deformation prescribed on the boundary. Again $\overset{*}{\underset{\sim}{u}}{}^{\alpha}$ represents the local fluctuations of the deformation.

Equation (2.7) defining the microdeformation considers the occurring deformations of a microelement with respect to its center of mass only. A modification of this model is indicated in Fig. 2(a,b). Here a reference frame is attached to the centre of mass of mass of each grain and the orientation of this frame to the fixed Eulerian frame is represented by a tensor $\underset{\approx}{O}^{\alpha}$ Hence, any arbitrary point in the undeformed state can be located by a vector $\underset{\sim}{X}^{\alpha}$ so that:

$$\underset{\sim}{X}^{\alpha} = \underset{\approx}{O}^{\alpha} \underset{\sim}{Y}^{\alpha} + \underset{\sim}{R}^{\alpha} , \tag{2.8}$$

expressing the orientation of the $\underset{\sim}{Y}^{\alpha}$ coordinates with respect to the $\underset{\sim}{X}^{\alpha}$ coordinates by $\underset{\approx}{O}^{\alpha}$ Similarly in the deformed state (see also Fig. 2b) this relation becomes:

$$\underset{\sim}{x}^{\alpha} = \underset{\approx}{o}^{\alpha} \underset{\sim}{y}^{\alpha} + \underset{\sim}{r}^{\alpha} . \tag{2.9}$$

It should be noted that in the present analysis majuscules refer to the undeformed and minuscules to the deformed state of the medium. This model and in particular the kinematics of deformation have been investigated by Haller [15] . The deformation process itself according to this model is deterministic and the quantities $\overset{\alpha}{\underset{\approx}{o}}$, $\overset{\alpha}{\underset{\sim}{r}}$ are only random due to the randomness of $\overset{\alpha}{\underset{\approx}{O}}$ and $\overset{\alpha}{\underset{\sim}{R}}$. It is readily seen that the microdeformation $\underset{\sim}{u}(\overset{\alpha}{\underset{\sim}{x}})$ in the proposed form of (2.8) and (2.9) is related to $\overset{\alpha}{\underset{\sim}{u}}$ as follows:

$$\underset{\sim}{u}(\overset{\alpha}{\underset{\sim}{x}}) = \overset{\alpha}{\underset{\sim}{x}} - \overset{\alpha}{\underset{\sim}{X}} = \overset{\alpha}{\underset{\sim}{u}} + \overset{\alpha}{\underset{\approx}{o}}\,\overset{\alpha}{\underset{\sim}{y}} - \overset{\alpha}{\underset{\sim}{O}}\,\overset{\alpha}{\underset{\approx}{Y}}. \tag{2.10}$$

In the present analysis this model, however, will not be further discussed and the simpler model outlined by (2.7) will be treated.

Following the definitions given in (2.6) and (2.7) the microdeformation can be expressed by:

$$\langle \overset{\alpha}{\underset{\sim}{u}} \rangle = \langle \overset{\alpha}{\underset{\sim}{r}} \rangle - \overset{\alpha}{\underset{\sim}{R}}\,; \qquad \overset{*\alpha}{\underset{\sim}{u}} = \overset{*\alpha}{\underset{\sim}{r}}, \tag{2.11}$$

in which $\overset{\alpha}{\underset{\sim}{u}}$ itself is a fluctuating quantity as shown in (2.7b). Equivalently, one may write for (2.11):

$$\overset{\circ\alpha}{\underset{\sim}{u}} = \overset{\circ\alpha}{\underset{\sim}{r}} - \overset{\circ\alpha}{\underset{\sim}{R}},$$

and since from (2.7a):

$$\left.\begin{aligned} \overset{\alpha}{\underset{\sim}{u}} &= \overset{\alpha}{\underset{\sim}{r}} - \overset{\alpha}{\underset{\sim}{R}} = \overset{\circ\alpha}{\underset{\sim}{r}} + \overset{*\alpha}{\underset{\sim}{r}} - \overset{\circ\alpha}{\underset{\sim}{R}} - \overset{*\alpha}{\underset{\sim}{R}}, \\ &= \overset{\circ}{\underset{\sim}{u}}(\overset{\circ\alpha}{\underset{\sim}{R}}) + \overset{*}{\underset{\sim}{u}}(\overset{\circ\alpha}{\underset{\sim}{R}}, \overset{*\alpha}{\underset{\sim}{R}}) \end{aligned}\right\} \tag{2.12}$$

It follows for small deformations, i.e., $\partial\overset{\alpha}{\underset{\sim}{u}}/\partial\overset{\alpha}{\underset{\sim}{r}} \approx \partial\overset{\alpha}{\underset{\sim}{u}}/\partial\overset{\alpha}{\underset{\sim}{R}}$ that:

$$\frac{\partial \overset{\alpha}{\underset{\sim}{u}}}{\partial \overset{\alpha}{\underset{\sim}{r}}} = \frac{\partial \langle \overset{\alpha}{\underset{\sim}{u}} \rangle}{\partial \overset{\alpha}{\underset{\sim}{r}}} + \frac{\partial \overset{*\alpha}{\underset{\sim}{u}}}{\partial \overset{\alpha}{\underset{\sim}{r}}}. \tag{2.13}$$

Alternatively:

$$\text{(2.14a)} \quad \frac{\partial \underset{\sim}{u}^{\alpha}}{\partial \underset{\sim}{r}^{\alpha}} = \frac{\partial \overset{\circ}{\underset{\sim}{u}}{}^{\alpha}}{\partial \underset{\sim}{r}^{\alpha}} + \frac{\partial \overset{*}{\underset{\sim}{u}}{}^{\alpha}}{\partial \underset{\sim}{r}^{\alpha}},$$

or:

$$\text{(2.14b)} \quad \frac{\partial \underset{\sim}{u}^{\alpha}}{\partial \underset{\sim}{r}^{\alpha}} = \left\langle \frac{\partial \underset{\sim}{u}^{\alpha}}{\partial \underset{\sim}{r}^{\alpha}} \right\rangle_{V^N} + \left(\frac{\partial \underset{\sim}{u}^{\alpha}}{\partial \underset{\sim}{r}^{\alpha}} \right)^{*}.$$

It is to be noted that $\left(\partial \underset{\sim}{u}^{\alpha} / \partial \underset{\sim}{r}^{\alpha}\right)^{*}$ is the fluctuation of $\partial \underset{\sim}{u}^{\alpha} / \partial \underset{\sim}{r}^{\alpha}$ and is <u>not</u> in general equal to $\partial \overset{*}{\underset{\sim}{u}}{}^{\alpha} / \partial \underset{\sim}{r}^{\alpha}$ which is <u>the gradient</u> of $\overset{*}{\underset{\sim}{u}}{}^{\alpha}$. Furthermore, $\partial \overset{\circ}{\underset{\sim}{u}}{}^{\alpha} / \partial \underset{\sim}{r}^{\alpha} \neq \langle \partial \underset{\sim}{u} / \partial \underset{\sim}{r} \rangle_{V^N}$. These deformations are defined for a particular microelement υ^{α} only as it moves with time. Hence, $\langle \partial \underset{\sim}{u} / \partial \underset{\sim}{r} \rangle_{V^N}$ is a spatial average within the mesoscopic domain for all deformations of υ^{α}, whilst $\left(\partial \underset{\sim}{u}^{\alpha} / \partial \underset{\sim}{r}^{\alpha}\right)^{*}$ is a random gradient within this region. In general $\left(\partial \underset{\sim}{u}^{\alpha} / \partial \underset{\sim}{r}^{\alpha}\right)^{*} \neq 0$ but $\left\langle \left(\partial \underset{\sim}{u}^{\alpha} / \partial \underset{\sim}{r}^{\alpha}\right)^{*} \right\rangle_{V^N} \equiv 0$. It follows further from relation (2.14a) that:

$$\text{(2.15)} \quad \begin{cases} \left\langle \dfrac{\partial \underset{\sim}{u}}{\partial \underset{\sim}{r}} \right\rangle_{V^N} = \left\langle \dfrac{\partial \underset{\sim}{u}}{\partial \underset{\sim}{r}} + \dfrac{\partial \overset{*}{\underset{\sim}{u}}}{\partial \underset{\sim}{r}} \right\rangle_{V^N}, \\[2ex] \qquad = \dfrac{\partial \overset{\circ}{\underset{\sim}{u}}}{\partial \overset{\circ}{\underset{\sim}{r}}} + \left\langle \dfrac{\partial \overset{*}{\underset{\sim}{u}}}{\partial \underset{\sim}{r}} \right\rangle_{V^N}, \end{cases}$$

and from (2.14b) that:

$$\left(\frac{\partial \underset{\sim}{u}^{\alpha}}{\partial \underset{\sim}{r}^{\alpha}}\right)^{*} = \frac{\partial \underset{\sim}{u}^{\alpha}}{\partial \underset{\sim}{r}^{\alpha}} - \left\langle \frac{\partial \underset{\sim}{u}}{\partial \underset{\sim}{r}} \right\rangle_{V^N} . \tag{2.16}$$

Using (2.14b) and introducing the symbol $\underset{\sim}{\Psi}^{\alpha}$ for the gradient of the microdeformation, then:

$$\underset{\sim}{\Psi}^{\alpha} = \langle \underset{\sim}{\Psi} \rangle_{V^N} - \overset{*}{\underset{\sim}{\Psi}}{}^{\alpha} . \tag{2.17}$$

Hence the microstrains and rotations can be expressed as:

$$\left.\begin{aligned} \underset{\sim}{\epsilon}^{\alpha} &= \frac{1}{2}\left(\underset{\sim}{\Psi}^{\alpha} + \underset{\sim}{\Psi}^{\alpha^T}\right) , \quad (a) \\ \underset{\sim}{\omega}^{\alpha} &= \frac{1}{2}\left(\underset{\sim}{\Psi}^{\alpha} \quad \underset{\sim}{\Psi}^{\alpha^T}\right) , \quad (b) \end{aligned}\right\} \tag{2.18}$$

where $\underset{\sim}{\Psi}^{\alpha T}$ is the transpose of $\underset{\sim}{\Psi}^{\alpha}$. On the boundary of the meso-scopic domain the macroscopic values are by definition:

$$\underset{\sim}{e} = \langle \underset{\sim}{\epsilon} \rangle_{V^N} \quad ; \quad \underset{\sim}{\Omega} = \langle \underset{\sim}{\omega} \rangle_{V^N} . \tag{2.19}$$

Considering strains only, then following (2.18a) it is seen that:

$$\underset{\sim}{e} = \frac{1}{2}\left(\langle \underset{\sim}{\Psi} \rangle_{V^N} + \langle \underset{\sim}{\Psi}^{T} \rangle_{V^N}\right) , \quad (a)$$

and:

$$\overset{*}{\underset{\sim}{\epsilon}}{}^{\alpha} = \frac{1}{2}\left(\overset{*}{\underset{\sim}{\Psi}}{}^{\alpha} + \overset{*}{\underset{\sim}{\Psi}}{}^{\alpha^T}\right) . \quad (b) \tag{2.20}$$

Hence using relation (2.15b) the macroscopic strain becomes:

(2.21a) (a) $$\underset{\sim}{e} = \frac{1}{2}\left[\frac{\partial \overset{\circ}{\underset{\sim}{u}}}{\partial \overset{\circ}{\underset{\sim}{r}}} + \left(\frac{\partial \overset{\circ}{\underset{\sim}{u}}}{\partial \overset{\circ}{\underset{\sim}{r}}}\right)^T\right] + \bar{\underset{\sim}{e}}$$

where:

(2.21b) (b) $$\bar{\underset{\sim}{e}} = \frac{1}{2}\left[\left\langle\frac{\partial \overset{*}{\underset{\sim}{u}}}{\partial \underset{\sim}{r}}\right\rangle_{V^N} + \left\langle\frac{\partial \overset{*}{\underset{\sim}{u}}}{\partial \underset{\sim}{\gamma}}\right\rangle_{V^N}^T\right]$$

It is seen from relation (2.21a) that a second order term occurs in this formulation of the macrostrains, which is not predicted from the conventional continuum theories. It is of interest to note, that the quantity $\bar{\underset{\sim}{e}}$ in (2.21b) could in certain cases be of the same order of magnitude as the first term in (2.21a).

Analogously to continuum theory it is possible to introduce a "relative displacement" tensor expressing the difference between the macroscopic displacement gradient and the micro-deformation such that:

(2.22) $$\underset{\sim}{\gamma} = \langle \underset{\sim}{\Psi} \rangle_{V^N} - \frac{\partial \overset{\circ}{\underset{\sim}{u}}}{\partial \overset{\circ}{\underset{\sim}{r}}} \quad ,$$

in which $\underset{\sim}{\gamma} \equiv \langle \overset{*}{\underset{\sim}{\Psi}} \rangle_{V^N}$ is considered to be caused by the fluctuating or residual stress field regarded here as a random tensor field. Then the quantity $\bar{\underset{\sim}{e}}$ is given by:

(2.23) $$\bar{\underset{\sim}{e}} = \frac{1}{2}\left(\underset{\sim}{\gamma} + \underset{\sim}{\gamma}^T\right) .$$

In a similar manner the rotations defined in (2.18b) can be written as:

$$\underset{\sim}{\omega}^{\alpha} = \underset{\sim}{\Omega} + \overset{*}{\underset{\sim}{\omega}}^{\alpha}, \tag{2.24}$$

and the macroscopic quantities then become:

$$\underset{\sim}{\Omega} \equiv \langle \underset{\sim}{\omega} \rangle_{V^N} \equiv \frac{1}{2}\left[\frac{\partial \overset{\circ}{\underset{\sim}{u}}}{\partial \overset{\circ}{\underset{\sim}{r}}} - \left(\frac{\partial \overset{\circ}{\underset{\sim}{u}}}{\partial \overset{\circ}{\underset{\sim}{r}}}\right)^{T}\right] + \frac{1}{2}\left(\left\langle \frac{\partial \overset{*}{\underset{\sim}{u}}}{\partial \underset{\sim}{\gamma}} \right\rangle_{V^N} - \left\langle \frac{\partial \overset{*}{\underset{\sim}{u}}}{\partial \underset{\sim}{r}} \right\rangle_{V^N}^{T}\right), \tag{2.25}$$

showing again a second term not given by the usual continuum mechanics formulation. Defining now another gradient which may be referred to as "macrogradient" such that:

$$\left.\begin{aligned} \underset{\sim}{\Lambda}^{\alpha} &= \frac{\partial \underset{\sim}{\Psi}^{\alpha}}{\partial \underset{\sim}{r}^{\alpha}} = \frac{\partial \langle \underset{\sim}{\Psi} \rangle_{V^N}}{\partial \underset{\sim}{r}^{\alpha}} + \frac{\partial \overset{*}{\underset{\sim}{\Psi}}^{\alpha}}{\partial \underset{\sim}{r}^{\alpha}}, \\ &= \frac{\partial^2 \overset{\circ}{\underset{\sim}{u}}^{\alpha}}{\partial \underset{\sim}{r}^{\alpha} \partial \underset{\sim}{r}^{\alpha}} + \frac{\partial^2 \overset{*}{\underset{\sim}{u}}^{\alpha}}{\partial \underset{\sim}{r}^{\alpha} \partial \underset{\sim}{r}^{\alpha}} . \end{aligned}\right\} \tag{2.26}$$

Using the permutation tensor $\underset{\sim}{\alpha}$ the compatibility relations for the micro and macro strain fields respectively can be expressed as follows:

$$\left.\begin{aligned} \underset{\sim}{\alpha}\underset{\sim}{\alpha} \frac{\partial^2 \underset{\sim}{\epsilon}^{\alpha}}{\partial \underset{\sim}{r}^{\alpha} \partial \underset{\sim}{r}^{\alpha}} &= 0 , \quad \underset{\sim}{\alpha} \frac{\partial \underset{\sim}{\Lambda}^{\alpha}}{\partial r^{\alpha}} = 0 , \quad (a) \\ \underset{\sim}{\alpha}\underset{\sim}{\alpha} \frac{\partial^2 \underset{\sim}{e}}{\partial \overset{\circ}{\underset{\sim}{r}}^{\alpha} \partial \underset{\sim}{r}^{\alpha}} &= 0 , \quad \underset{\sim}{\alpha} \frac{\partial \langle \Lambda \rangle_{V^N}}{\partial \overset{\circ}{\underset{\sim}{r}}} = 0 . \quad (b) \end{aligned}\right\} \tag{2.27}$$

The proof of the above relations is readily obtained. It can be shown further that by using the above microdeformations, specific boundary conditions relevant to the boundary of the mesoscopic domain and a material functional representing the constitutive relations for the medium, the corresponding field equations can be formulated.

3. Constitutive Relations

In the foregoing discussion the significant quantities during the deformation of a structured medium have been considered as random quantities. In particular the induced stress field within the mesoscopic domain is a random tensor field. For its statistical description an $6N$-dimensional distribution is required for the set values:

$$\tilde{\xi}^{\alpha} = \tilde{\xi}(\tilde{r}^{\alpha}) \; ; \; \alpha = 1, \ldots, N \, . \tag{3.1}$$

This field will be specified within the region V^N to every finite set of points r^{α}, $\alpha, \ldots, N$. The $6N$-dimensional distribution will be of the form:

$$P\left\{\xi^{1}_{11}, \xi^{1}_{12}, \cdots, \xi^{1}_{33}, \xi^{2}_{11}, \cdots, \xi^{N}_{33}\right\} = P\left\{\tilde{\xi}(\tilde{r}^{\alpha})\right\}, \tag{3.2}$$

for the random variable $\tilde{\xi}^{\alpha}$ and where its density function must satisfy certain conditions [16]. From this distribution one can express higher moments of this quantity, but for most practical problems an approximation by using only second moments will be sufficient. Hence, by employing correlation theory, the random stress tensor field can be represented by its correlation tensor which in the present case may be written as:

$$\underset{\sim}{R} = \left\langle \overset{*}{\underset{\approx}{\xi}}(\underset{\sim}{r}^{\alpha_1})\, \overset{*}{\underset{\approx}{\xi}}(\underset{\sim}{r}^{\alpha_2}) \right\rangle_{V^N} , \qquad (3.3)$$

involving the fluctuating components of the microstress at positions $\underset{\sim}{r}^{\alpha_1}, \underset{\sim}{r}^{\alpha_2}$. From a purely geometric point of view by considering the parameter φ of the material functional $\underset{\sim}{\mathfrak{F}}^{\tau}$ inequation (1.1) only, the equilibrium of the stress field for any $\varphi \geqslant 0$ can be expressed by:

$$\left.\begin{aligned} &\frac{\partial \underset{\approx}{\xi}(\underset{\sim}{r}^{\alpha_1} + \underset{\sim}{\varphi})}{\partial \underset{\sim}{r}^{\alpha_1}} = \frac{\partial \underset{\approx}{\xi}(\underset{\sim}{r}^{\alpha_1} + \underset{\sim}{\varphi})}{\partial \underset{\sim}{\varphi}} = 0, \\ &\text{and also:} \\ &\frac{\partial \underset{\sim}{R}}{\partial \underset{\sim}{\varphi}} = 0 \quad \text{at} \quad \underset{\sim}{\varphi} = 0 . \end{aligned}\right\} \qquad (3.4)$$

Equation (3.3) concerning the stress correlation tensor in the case of a linear response behaviour of a structured medium has been studied in some detail by Helbawi [13]. It may be noted that a simplification of the analysis can be achieved by assuming that the field $\underset{\sim}{\xi}$ is statistically isotropic. In this case the fundamental moments depend only on $(N-1)$ vectors that determine the configurations of the microelements at $\underset{\sim}{r}^{\alpha}$. Then $\langle \underset{\sim}{\xi} \rangle_{N^{V}}$ is constant and equal to the macroscopic stress tensor $\underset{\sim}{\sigma}$ on the boundary of the mesoscopic domain.

In representing the structure of the medium by a discrete ensemble of microelements υ^{α} that are subjected to a stochastic deformation process, it has so far been shown that the central problem is the determination of the probability distributions of the quantities involved. In particular, the evolution of the probability density function, in the case of a more general time dependent deformation, will require a knowledge of the operator A^{τ} in the material functional $\underset{\sim}{\mathfrak{F}}$. It has been shown in previous work [17] that it is convenient for the study of this operator to utilize system theory.

Thus, considering first the simple case of a microelement υ^{α} that is subjected to a stimulus $\underset{\sim}{\xi}^{\alpha}$ then:

(3.5) $$\underset{\sim}{\epsilon}^{\alpha} = \underset{\sim}{b}\, \underset{\sim}{\xi}^{\alpha}$$

where $\underset{\sim}{b}$ is the "local" transform operator, which in the very restricted linearized case is a fourth order tensor representing the elastic property of the medium. If time-dependency of the deformation is introduced, then relation (3.5) becomes:

(3.6) $$\underset{\sim}{\epsilon}^{\alpha}(t) = \underset{\sim}{b}^{\tau}\, \underset{\sim}{\xi}^{\alpha}(\tau)$$

where $\underset{\sim}{b}^{\tau}$ is a transform operator representing the transfer from the stimulus $\underset{\sim}{\xi}^{\alpha}$ to the response $\underset{\sim}{\epsilon}^{\alpha}$. In terms of a "delayed impulse" function, the stimulus can also be written as:

$$\underset{\sim}{\xi}^{\alpha}(t) = \int_0^t \underset{\sim}{\xi}^{\alpha}(\tau)\, \delta(t-\tau)\, d\tau. \tag{3.7}$$

and the response at the αth microelement as:

$$\underset{\sim}{\epsilon}^{\alpha}(t) = \int_0^t \underset{\sim}{g}^{\alpha}(t, \tau)\, \underset{\sim}{\xi}^{\alpha}(\tau)\, d\tau, \tag{3.8}$$

where $\delta(t-\tau)$ is the Dirac-delta function and $\underset{\sim}{g}^{\alpha}(t-\tau)$ the impulse transfer function of the αth element. It represents the response of the αth microelement at time t to an excitation $\delta(t-\tau)$ which is a unit pulse at time t. Further:

$$\underset{\sim}{g}^{\alpha}(t, \tau) = \underset{\sim}{b}^{\tau}\delta(t-\tau); \quad \underset{\sim}{g}^{\alpha}(t-\tau) = 0 \text{ for } \tau > t. \tag{3.9}$$

Hence, in this formulation the operator $\underset{\sim}{b}^{\tau}$ of the material functional $\underset{\sim}{\mathfrak{F}}^{\tau}$ becomes a stochastic integral operator. If the response of an array of microelements in the mesoscopic domain subjected to a 'single' stochastic input $\underset{\sim}{\xi}^{\alpha}$ are considered, given by the set $\underset{\sim}{\epsilon}^{1\alpha}, \ldots, \underset{\sim}{\epsilon}^{M\alpha}$, $\alpha = 1, \ldots, N$, then:

$$\underset{\sim}{\epsilon}^{\beta}(t) = \sum_{\alpha=1}^{N} \underset{\sim}{\epsilon}^{\beta\alpha}(t); \beta = 1, \cdots, M, \tag{3.10}$$

and

$$\underset{\sim}{\epsilon}^{\beta\alpha}(t) = \underset{\sim}{b}_{\tau}^{\beta\alpha}\, \underset{\sim}{\xi}^{\alpha}(\tau), \tag{3.11}$$

in which $\underset{\sim}{b}_{\tau}^{\beta\alpha}$ is now a space-wise stationary transform operator representing the material properties. The superscript of $\underset{\sim}{b}$ then refers to φ of $\underset{\sim}{\mathfrak{F}}^{\tau}$ indicating the interaction between microelements α at the input and β at the output of the ensemble. In terms of the vectorial distance φ can be expressed as before by $\varphi^{\alpha\beta} = |\underset{\sim}{r}^{\alpha} - \underset{\sim}{r}^{\beta}|$. Following (3.8) the relation in the case of an array of microelements becomes:

$$\underset{\sim}{\epsilon}^{\beta\alpha}(t) = \int_0^t \underset{\sim}{g}^{\beta\alpha}(t, \tau)\ \underset{\sim}{\xi}^{\alpha}(\tau)\, d\tau, \tag{3.12}$$

and

$$\underset{\sim}{\epsilon}^{\beta}(t) = \sum_{\alpha=1}^{N} \int_0^t \underset{\sim}{g}^{\beta\alpha}(t, \tau)\ \underset{\sim}{\xi}^{\alpha}(\tau)\, d\tau. \tag{3.13}$$

An equation of state based on a multi-dimensional system can now be formulated. A detailed treatment of stochastic dynamic models is given in reference [17] where the stimuli $\underset{\sim}{\xi}^{\alpha}$ and responses $\underset{\sim}{\epsilon}^{\alpha}$ associated with each microelement are taken with reference to a fixed Eulerian coordinate system at any time $0 \leq \tau \leq t$ and where for simplicity of the analysis the position of the element is considered as that of its center of mass.

The above discussion indicates the treatment of the more general time-dependent deformation process by using stochastic dynamic models to establish constitutive relations. With regard to the evolution of the transform operator $\underset{\sim}{b}^{\tau}$, during a time-dependent deformation, reference is made to previous work [18]. In the latter it has been hypothesized that this operator will assume certain values in the probability space in such a manner that the evolution with time follows a Markov process. In this context this evolution has been considered as a space-wise discontinuous, but time-wise continuous process, i.e. a 'discontinuous Markov process'. On this basis a distribution function for the operator involving a "transition matrix' has been obtained.. This matrix representing the transition probabilities from one state of the deformation to an adjacent state can be established from two successive experimental observations [19]. Finally employing the theorems of statistical mechanics and stochastic operators of random theory, a transfer function valid for the mesoscopic domain can be formed and the expected values of stresses and strains related to each other. However, relations of this type are restricted in the sense that linear operators are used and that the deviations from thermodynamic equilibrium are small.

4. Functional Analysis

Using the formulation of microdeformations presented in section 2 and the approach outlined in section 3, the corresponding field equations can be established. However, the representation of the field equations refers to a quasi-continuum since the mass density appears in the probability distributions and the deformations in terms of local quantities. Hence in order to obtain a transition from a discrete situation to a continuous one for which the well-known forms of continuum theory apply, the phase-space can be introduced as a function space. The latter then becomes a topological linear vector space whereby the deformation process is described by a finite number of functions rather than by a finite number of parameters. In this context it should be mentioned that such an approach is discussed in reference [20] dealing with the thermodynamics of the deformation process. Considering now the structured medium as a dynamical system with a finite number of degrees of freedom and introducing generalized coordinates $\underset{\sim}{q}^{\alpha}$, $\alpha = 1,\dots,N$ and momenta $\underset{\sim}{p}^{\alpha}$, $\alpha = 1,\dots,N$, the probability density functions in the Γ-space can be expressed by:

$$ {}^{\Gamma}f = {}^{\Gamma}f\,(\underset{\sim}{q}^{\alpha}, \underset{\sim}{p}^{\alpha}, t)\,, \tag{4.1} $$

which is subject to the normalizing condition:

$$ \int_{\Gamma} {}^{\Gamma}f\, d\Gamma = 1 \;; \quad d\Gamma = \prod_{i=1}^{3} dq_i^{1} \cdots dq_i^{N}\, dp_i^{1} \cdots dp_i^{N}\,, \tag{4.2} $$

or equivalently ${}^{\Gamma}f d\Gamma$ being the probability of the center of mass of each microelement or grain having the coordinates $\underset{\sim}{q}^{\alpha} = \underset{\sim}{r}^{\alpha}$, $\underset{\sim}{p}^{\alpha} = \dot{\underset{\sim}{r}}^{\alpha} \rho^{\alpha}/\rho$ in the range $d\Gamma$ at the instant t . The factor ρ^{α}/ρ is the ratio of the mass density of a microelement to the mass density averaged over the entire mesoscopic domain (where $\rho = \frac{1}{N}\sum_{\alpha=1}^{N} \rho^{\alpha}$).

These density functions satisfy the Liouville relations as follows:

$$\frac{\partial\,^{\Gamma}f}{\partial t} + \sum_{i=1}^{3}\sum_{\alpha=1}^{N}\left[\frac{\partial\,^{\Gamma}f}{\partial q_i^{\alpha}}\;\frac{\partial q_i^{\alpha}}{dt} + \frac{\partial\,^{\Gamma}f}{\partial p_i^{\alpha}}\;\frac{d p_i^{\alpha}}{dt}\right] = 0 \,. \qquad (4.3)$$

When a more general deformation process is considered, i.e., where the microelements themselves are regarded as deformable, a functional ${}_{\Gamma}\mathfrak{F}$ containing one or more additional dynamic variables can be introduced. Thus:

$${}_{\Gamma}\mathfrak{F} = {}_{\Gamma}\mathfrak{F}\left(\underset{\sim}{q}^{\alpha}, \underset{\sim}{p}^{\alpha}, \underset{\sim}{x}^{\alpha}; t\right) \,, \qquad (4.4)$$

which would then represent the model indicated in Fig. 2. The expected value of this functional can be written as:

$$\langle {}_{\Gamma}\mathfrak{F} \rangle_{\Gamma} = \int_{\Gamma} {}_{\Gamma}\mathfrak{F}\,^{\Gamma}f\, d\Gamma \;; \qquad \underset{\sim}{x}^{\alpha} \neq \underset{\sim}{r}^{\alpha} \,, \qquad (4.5)$$

and where for its existence the probability distribution is assumed such that:

$$0 < \int_{\Gamma} {}^{\Gamma}f\, \delta\left(\underset{\sim}{q}^{\alpha} - \underset{\sim}{x}^{\alpha}\right) d\Gamma \leqslant 1 \,, \qquad (4.6)$$

in which the choice of the generalized coordinates $\underset{\sim}{q}^{\alpha}$ are restricted to the Eulerian frame defined for each individual microelement. The function $\delta\left(\underset{\sim}{q}^{\alpha} - \underset{\sim}{x}^{\alpha}\right)$ in the integrant of (4.6) is the 3-dimensional Dirac-delta function. The transformation from the Γ-space to the physical space of the medium V^{N} can be written as:

$$(4.7)\qquad \langle \underset{\sim}{\mathfrak{F}}^{\tau} \rangle_{V^N} \Longrightarrow \int_{\Gamma} \mathfrak{F}\, {}^{\Gamma}f\ \delta(\underset{\sim}{q}^{\alpha} - \underset{\sim}{x}^{\alpha})\, d\Gamma; \qquad \underset{\sim}{x}^{\alpha} = \underset{\sim}{r}^{\alpha}$$

where $\underset{\sim}{\mathfrak{F}}^{\tau}$ is the material functional defined by (1.1) but considered now in terms of its distribution function within the mesoscopic domain. It is seen from relations (4.5) and (4.7) that the expected value of this functional $\underset{\sim}{\mathfrak{F}}^{\tau}$ will depend on the chosen model. In the case of $\underset{\sim}{x}^{\alpha} \neq \underset{\sim}{r}^{\alpha}$ equation (4.5) will hold, whilst for $\underset{\sim}{x}^{\alpha} = \underset{\sim}{r}^{\alpha}$ equation (4.7) will apply. Denoting the deformation in the Γ-space by $\hat{\underset{\sim}{u}}^{\alpha} = \hat{\underset{\sim}{u}}(\underset{\sim}{q}^{\alpha}, \underset{\sim}{p}^{\alpha}, \underset{\sim}{x}^{\alpha}, t)$, the average can be written in accordance with (4.5) and (4.7) respectively, as:

$$(4.8)\qquad \underset{\sim}{u}^{\alpha}\Big|_{\underset{\sim}{x}^{\alpha} \neq \underset{\sim}{r}^{\alpha}} = \langle \hat{\underset{\sim}{u}}^{\alpha} \rangle\Big|_{\underset{\sim}{x}^{\alpha} \neq \underset{\sim}{r}^{\alpha}} = \int_{\Gamma} {}^{\Gamma}f\, \hat{\underset{\sim}{u}}^{\alpha}\, d\Gamma ,$$

and

$$(4.9)\qquad \overset{\circ}{\underset{\sim}{u}}{}^{\alpha}\Big|_{\underset{\sim}{x}^{\alpha} = \underset{\sim}{r}^{\alpha}} = \langle \hat{\underset{\sim}{u}}^{\alpha} \rangle_{\underset{\sim}{x}^{\alpha} = \underset{\sim}{r}^{\alpha}} = \int_{\Gamma} {}^{\Gamma}f\, \hat{\underset{\sim}{u}}^{\alpha}\, \delta(\underset{\sim}{q}^{\alpha} - \underset{\sim}{x}^{\alpha})\, d\Gamma ,$$

in which as previously discussed $\overset{\circ}{\underset{\sim}{u}}{}^{\alpha}$ represents the deformation estimated from the boundary conditions on the mesoscopic domain and $\underset{\sim}{u}^{\alpha}$ would account for the deformation, if the microelements are considered as deformable. Hence the deformation

$$(4.10)\qquad \underset{\sim}{u}^{\alpha} + \overset{\circ}{\underset{\sim}{u}}{}^{\alpha} = \underset{\sim}{u}(\underset{\sim}{x}^{\alpha}),$$

is the deformation at any arbitrary point α in the physical space.

It is to be noted that although these averages are obtained in the Γ-space they correspond due to the transformation relation (4.7) to specific microelements in the physical space. Hence all elements are mapped onto a space in which each element is a projection of the Γ-space onto a three-dimensional Euclidean space. However, in order to proceed with the formulation, a proper metric between the function space $\left[\Gamma, {}_{\Gamma}\mathfrak{F}(\underset{\sim}{q}, \underset{\sim}{p}, \underset{\sim}{x}, t)\, \mu\right]$ and the physical space or its probability space $\left[V^{N}, \underset{\sim}{\mathfrak{F}}^{\tau}, \mu\right]$ must be established. This can be achieved by assigning the probability measure μ to these spaces. The latter is not unique in general unless certain restrictions are made on the distributing of the functional $\mathfrak{F}^{\tau}$. Thus, for instance, it may be assumed that the distribution is Gaussian and that it is a limiting case through the application of the "central limit theorem" of probability theory. This will be the case for a system with a large number of degrees of freedom within the mesoscopic domain. In this manner the field equations in the function space analogous to those in the physical space can be expressed using (4.9) as follows:

$$ {}_{V^N}\underset{\sim}{L}\left(\langle \underset{\sim}{u}^{\alpha} \rangle\right) \Longrightarrow \langle {}_{\Gamma}\underset{\sim}{L}(\hat{\underset{\sim}{u}}^{\alpha}) \rangle_{\Gamma} = \int_{\Gamma} {}_{\Gamma}\underset{\sim}{L}(\hat{\underset{\sim}{u}}^{\alpha})\, {}^{\Gamma}f\, \delta(\underset{\sim}{q}^{\alpha} - \underset{\sim}{x}^{\alpha})\, d\Gamma, \tag{4.11} $$

where ${}_{\Gamma}\underset{\sim}{L}(\hat{\underset{\sim}{u}}^{\alpha})$ represents a linear operator:

(4.12) $$ {}_{\Gamma}\underset{\sim}{L}(\hat{\underset{\sim}{u}}^{\alpha}) = \rho^{\alpha}\underset{\sim}{1}\,\frac{\partial^{2}\hat{\underset{\sim}{u}}^{\alpha}}{\partial t^{2}} $$

Again if the microelement is considered as deformable another operator $\{{}_{\Gamma}\underset{\sim}{Q}\}(\hat{\underset{\sim}{u}}^{\alpha})$ will have to be used. The latter can be written as shown in reference (12) as follows:

(4.13) $$ \{{}_{\Gamma}\underset{\sim}{Q}\}(\hat{\underset{\sim}{u}}^{\alpha}) = \Big[\rho^{\alpha}\underset{\sim}{1}\,\frac{\partial^{2}\hat{\underset{\sim}{u}}^{\alpha}}{\partial t^{2}} - \mu^{\alpha}\underset{\sim}{1}\,\Delta^{2}\underset{\sim}{u}^{\alpha} - (\mu^{\alpha}+\lambda^{\alpha})\,\underset{\sim}{\Delta}(\underset{\sim}{\Delta}\cdot\hat{\underset{\sim}{u}}^{\alpha})\Big] = {}_{\Gamma}\underset{\sim}{F}^{\alpha} $$

in which the coefficients $\mu^{\alpha}, \lambda^{\alpha}$ correspond to the conventional material constants at the α th microelement and ${}_{\Gamma}\underset{\sim}{F}^{\alpha}$ denotes a generalized force. It is difficult, however, to define such a force in the Γ-space, but it can be obtained by using the transformation rule from before, i.e.:

(4.14) $$ \Big\langle\{{}_{\Gamma}\underset{\sim}{Q}\}(\hat{\underset{\sim}{u}}^{\alpha})\Big\rangle_{\Gamma} = \langle{}_{\Gamma}\underset{\sim}{F}^{\alpha}\rangle \equiv \underset{\sim}{F}(\underset{\sim}{x},\underset{\sim}{t}) , $$

in which now $\underset{\sim}{F}(\underset{\sim}{x},\underset{\sim}{t})$ is the given force field acting on the mesoscopic region. If both relations (4.13) and (4.14) are taken into account in the case of a more general deformation, then the left-hand side of (4.14) will read:

(4.15) $$ {}_{\Gamma}\underset{\sim}{L}\,\hat{\underset{\sim}{u}}^{\alpha} + {}_{\Gamma}\bar{\underset{\sim}{Q}}'\,\hat{\underset{\sim}{u}}^{\alpha} = {}_{\Gamma}\underset{\sim}{F}^{\alpha} ; \quad {}_{\Gamma}\bar{\underset{\sim}{Q}}' = \Big\langle\{{}_{\Gamma}\underset{\sim}{Q}'\}\Big\rangle_{\Gamma} $$

It is seen from the above discussion that in the function space approach the conventional field equations can be used with the modification due to the presence of the operators ${}_{\Gamma}\underset{\sim}{L}$ and ${}_{\Gamma}\bar{\underset{\sim}{Q}}'$. For a more complete formulation even for the simple model used, interaction effects manifested in terms of surface potential, etc., would have to be included. Such effects will then change the form of the operator ${}_{\Gamma}\bar{Q}'$, since it will no longer remain independent of the Γ-space coordinates, when it is considered in that space.

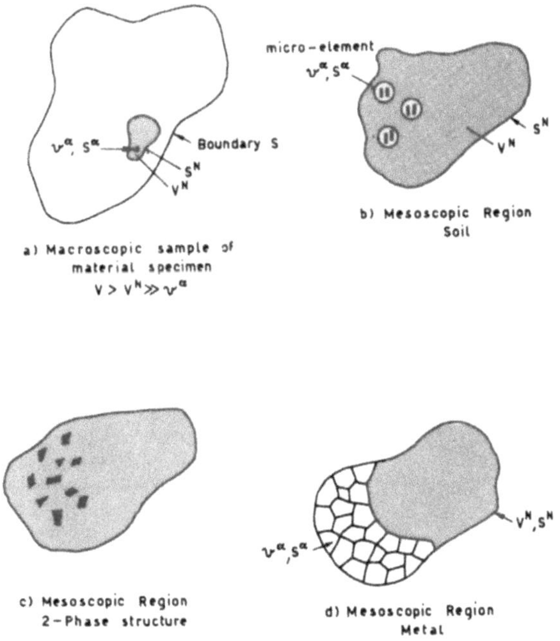

Fig. 1. Mesoscopic Regions

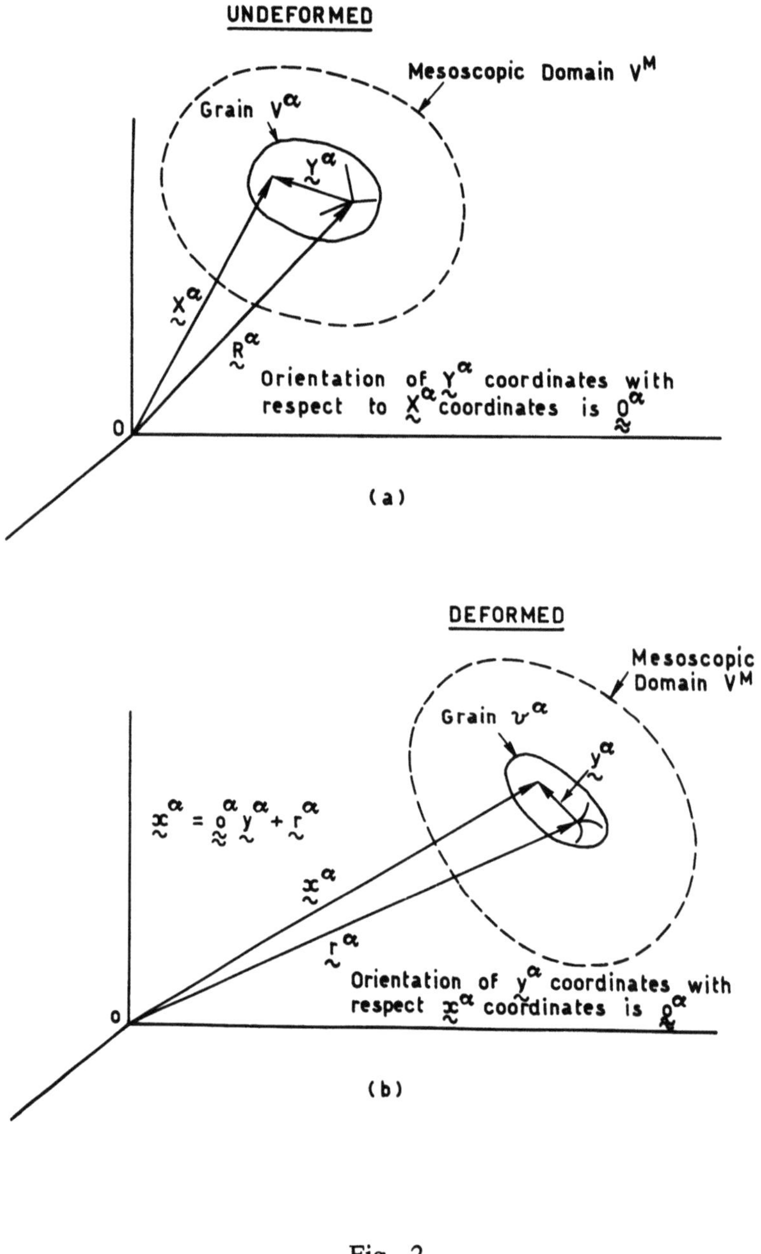
UNDEFORMED
Mesoscopic Domain V^M
Grain V^α
Y^α
X^α
R^α
Orientation of Y^α coordinates with respect to X^α coordinates is O^α
O
(a)
DEFORMED
Mesoscopic Domain V^M
Grain v^α
y^α
$x^\alpha = o^\alpha y^\alpha + r^\alpha$
x^α
r^α
Orientation of y^α coordinates with respect x^α coordinates is o^α
O
(b)

Fig. 2.

REFERENCES

[1] E.& F. COSSERAT, "Theorie des Corps Deformables" A. Hermann & Fils, Paris (1909).

[2] P. DUHEM, "Arm. Ecole Norm" 10, 187, (1893).

[3] C. TRUESDELL & R. TOUPIN, "The Classical Field Theories". Handbuch der Physik, 111/1, Springer-Verlag, Germany, (1960).

[4] R. D. MINDLIN, "Micro-structure in Linear Elasticity" Arch. Rational Mech. Anal., 16, 51-78, (1964).

[5] J. L. ERICKSEN & C. TRUESDELL, "Exact Theory of Stress and Strain in Rods and Shells'. Arch. Rational Mech. Anal., 1, 295-323, (1958).

[6] R. A. TOUPIN, "Elastic Materials with Couple Stresses" Arch. Rational Mech. Anal., 11, 385-414, (1962).

[7] A. C. ERINGEN & E.S. SUHUBI, 'Nonlinear Theory of Micro-stucture - 1". Int. J. Engrg. Sci., 2, 189-203, (1964).

[8] A. E. GREEN & R. S. RIVLIN, "Multipolar Continuum Mechanics", Arch. Rational Mech. Anal., 17, 113-147 (1964).

[9] D. R. AXELRAD, 'Stochastic Analysis of Flow of Micro-inhomogeneous Media', 5th U.S. Nat. Congr. Appl. Mech., University of Minnesota, (1966).

[10] D. R. AXELRAD, "Stochastic Analysis of Flow in Two-Phase Media", 5th Int. Congr. Rheology, Kyoto University, Kyoto, Japan, (1968).

[11] D. R. AXELRAD & R. N. YONG, "An Isothermal Flow for a heterogeneous medium" in Modern Developments

in the Mechanics of Continua, Academic Press, New York, (1966).

[12] D. R. AXELRAD & L. G. JAEGER, "Random Theory of Deformation in Heterogeneous Media", Int. Conf. Solid Mechs. (Southampton), England (1969).

[13] S. EL HELBAWI, 'The effects of Dislocations on the Linear Response of Elastic Heterogeneous Solids', M. Eng. Thesis, McGill University, (1971).

[14] A. M. YAGLOM, 'Theory of Stationary Random Functions" Prentice-Hall Inc., (1962).

[15] J. K. HALLER, "Kinematics of Deformation of Inhomogeneous Solids", M. Eng. Thesis, McGill University, (1971).

[16] V. A. LOMAKIN, "On the Theory of Deformation....." P.M.M., 30, 875-881, (1966) or, J. Appl. Math. Mech., 30, 1035-1042, (1966).

[17] D. R. AXELRAD & J. W. PROVAN, "Stochastic Models of Relaxation Phenomena", Rheological Acta, 10, 330-335, (1971).

[18] D. R. AXELRAD, "Rheology of Structured Media", Archum. Mech. Stosow., 23, 1, 131-140 (1971).

[19] J. KALOUSEK, "Stress Holographic Interferometry", M. Eng. Thesis, McGill University, (1971).

[20] D. R. AXELRAD & J. W. PROVAN, 'Thermodynamics of Deformation in Structured Media", In this volume.

D. R. AXELRAD – J. W. PROVAN

THERMODYNAMICS OF DEFORMATION IN STRUCTURED MEDIA

P R E F A C E

Using the concepts of the random theory previously introduced, the thermodynamics of deformation of structured media deal first with the formulation of an evolution criterion. Based on this criterion, a "Master equation" is then developed which governs the irreversible evolution of the deformation.

Udine, July 1971

D.R. Axelrad

1. Introduction

With the advent of an era which is making increasing industrial and scientific use of composite materials and materials whose properties on the crystalline, polycrystalline or fibrous level are crucial to their response characteristics, not enough is known concerning the thermodynamics of deformation in such materials. For such materials, the thermodynamics applied to equilibrium situations, alternatively known as thermostatics, is well established with its subject matter admirably verified through the use of statistical mechanics. The situation is quite different, however, in the case of non-equilibrium thermodynamics, which may be referred to as the thermodynamics of processes with the implication that it applies to irreversible phenomena.

During the deformation of a structured medium non-equilibrium states are attained and it becomes impossible to describe the deformation process in terms of either phenomenological or reversible laws pertaining to the thermodynamic state of the constituents forming the structure of the medium. It has been shown in previous work, for example by D.R. Axelrad and L. G. Jaeger [1], that in the formulation of the deformation field a random theory approach may be used, which is based on statistical mechanics and probability theory. In this theory the relevant quantities are considered as stochastic variables and, to a first approximation, correlation theory is employed to find their distributions by equating their expected values and second moments to those macroscopically observed on the boundary of a specified domain. These specific domains, called mesoscopic regions, contain a large number of microelements so that the principles of statistical mechanics may be applied. These concepts are also utilized in the present work.

From a thermodynamics point of view, a distinct feature of the approach is that an evolution criterion for the process of deformation in structured media is sought. Such a criterion is derived from the statistical theory of non-isolated or open thermodynamic systems since in all cases of interest the boundary conditions are such that an exchange of energy, in the form of heat, work or matter, must be allowed for in the formulation. Furthermore, it is well known that while the macroscopic formulation distinguishes between past and future of a thermodynamic process, the microscopic one does not. It is evident therefore that in order to close the gap between both representations either additional assumptions must be made or more information on such thermodynamic processes must be obtained. The formulation in classical statistical mechanics tries to close this gap in terms of an approximate relation, referred to as a master equation, and it is one of the aims of the present investigation to formulate such a master equation for structured media. However, before dealing with these propositions it is necessary to clarify the concept of phase densities in the μ -space and Γ -space pertaining to the problem and this is carried out in the next section.

2. Phase Densities

In reviewing briefly the concepts and definitions introduced in previous studies, for example by D.R. Axelrad [2] and J.W. Provan [3], a solid structured medium is considered to consist of an aggregate of crystals of "microelements" the material properties and response characteristics of which are governed by the laws of continuum mechanics. This explicity excludes, therefore, dislocation effects and interfacial potentials, etc. Each individual microelement α is considered as an open thermodynamic system whose state is represented by the set of state variables $\{\nu_i^\alpha\}$, $i = 1, \ldots, r; \alpha = 1, \ldots, N$, where r is the number of state parameters and N the number of microelements in a specified domain. For convenience this set may be regarded as an r -dimensional vector $\underset{\sim}{\nu}^\alpha$. Due to the inherent inhomogeneities and the variation in geometries and physical properties of the structural medium, the state parameter vector $\underset{\sim}{\nu}^\alpha$ is considered to be a "stochastic variable". The number N designates a large number of microelements contained in a specified domain of the material sample, which is referred to as a "mesoscopic domain'. Hence, the notion is introduced here that the macroscopic body, which is analogously termed a "macroscopic domain", consists of a large and unspecified number of mesoscopic domains. On the boundary of the macroscopic domain macroscopically observable quantities, such as stress and strain, etc., may be specified so that the average or expected values of the state parameters can be assessed for each mesoscopic domain by using phenomenological laws.

The thermodynamic state of the α th microelement may be represented in a single point in an r -dimensional "μ -space' with coordinates ν_i . The physical state of each mesoscopic domain therefore forms a cluster of N points in this μ -space. However, since it is impossible to determine by experimental observation the precise values of the r state variables for each microelement, any observation, which is necessarily a macroscopic one, can only supply information as to whether the α-th element is in the thermodynamic state whose range is:

$$ {}^{\ni}\underset{\sim}{\nu} \leq \overset{\alpha}{\underset{\sim}{\nu}} < {}^{\ni}\underset{\sim}{\nu} + \Delta \underset{\sim}{\nu}. \tag{2.1} $$

in which the letter "$\ni$" prior to the state vector designates a particular range in μ -space and $\Delta \underset{\sim}{\nu}$ represents the accuracy to which the experimental observations may be made. Hence, the μ -space is divided into a set of "cells" $\left\{ {}^{\ni}\omega \right\}$, each corresponding to the inaccuracies $\Delta \nu$, occupying a volume:

$$^{\ni}\omega = \prod_{i=1}^{r} \Delta \nu_i \tag{2.2}$$

in μ -space. Following P. and T. Ehrenfest [4], it is assumed that these cells are small compared to the measurable dimensions, but large enough to contain the representation of the state of a large number of microelements.

The physical state of a mesoscopic domain is described by N_r state variables ν_i^{α}, $i = 1, \ldots, r; \alpha = 1, \ldots, N$. This state may be represented by a single point on a N_r -dimensional "Γ-space" with the thermodynamic state of the complete macroscopic domain represented by a cluster of such points. The density of points in this Γ-space is a "fine grained" density ρ. Since each mesoscopic domain contains a large representative number of microelements, the observed state of each mesodomain may be described by the set of occupation numbers $\{^{\ni}n\}$ which designates the number of microelements whose state appears in the cells $^{\ni}\omega$ in μ -space. The set $\{^{\ni}n\}$ and its time evolution describe the nature and response characteristics of the specific material under investigation. Hence, its knowledge constitutes the knowledge of a "material functional" $\underset{\sim}{\mathfrak{F}}^{\tau}$ discussed by D.R. Axelrad [2]. Corresponding to the cell volume $^{\ni}\omega$ in μ -space, with occupation numbers $\{^{\ni}n\}$ there is a volume in Γ -space which is given by:

$$\Delta\Gamma = \prod_{\ni} {}^{\ni}\omega^{{}^{\ni}n} \; ; \; \sum_{\ni} {}^{\ni}n = N . \tag{2.3}$$

However, since the permutation of microelements whose state is described by the cells ${}^{\ni}\omega$ of μ -space does not alter the physical state of the mesoscopic domain there are:

$$^{n}\Lambda = \frac{N!}{\prod_{\ni} {}^{\ni}n!} , \tag{2.4}$$

$\Delta\Gamma$ volumes in Γ -space corresponding to this state. The superscript "n" prior to Λ indicates its dependence on the set $\{{}^{\ni}n\}$. It is seen therefore that each set $\{{}^{\ni}n\}$ defines a neighbourhood in Γ -space, called a "star" ${}^{n}\Omega$ whose measure may be expressed by:

$$M({}^{n}\Omega) = {}^{n}\Lambda\,\Delta\Gamma = \frac{N!}{\prod_{\ni} {}^{\ni}n!} \prod_{\ni} {}^{\ni}\omega^{{}^{\ni}n} . \tag{2.5}$$

Hence, by constructing these stars in Γ -space it is possible to describe the state of each mesoscopic domain to within a certain accuracy. This thus constitutes a "course graining" of Γ -space. Consequently, while precise values of the fine grained density ρ cannot be observed, the statistical weight of each star ${}^{n}\Omega$, corresponding to the integral:

$$\int_{{}^{n}\Omega} \rho \, d\Gamma ,$$

may be assessed. Using this quantity, the probability ${}^{n}f$ of finding the thermodynamic state of a specific mesoscopic domain at a point of ${}^{n}\Omega$ is given by:

$$ {}^{n}f = \frac{\int_{{}^{n}\Omega} \rho \, d\Gamma}{M({}^{n}\Omega)} . \tag{2.6}$$

Hence, a course grained density of the stochastic state variables ν_i^d in Γ -space is obtained. Furthermore, it should be noted that since ${}^{n}f$ represents the average value of the fine grained density taken over ${}^{n}\Omega$ it is constant over this region and satisfies:

$$\sum_{n} {}^{n}f \, M({}^{n}\Omega) = 1 \quad \text{or} \quad \int_{\Gamma} {}^{n}f \, d\Gamma = 1 . \tag{2.7}$$

The "probability" of finding the state of the mesoscopic region within the state ${}^{n}\Omega$ is therefore given by:

$$ {}^{n}P = {}^{n}f \, M({}^{n}\Omega) . \tag{2.8}$$

By introducing these phase densities and their transformations from the physical space through the μ-space to the Γ -space, the time evolution of the macroscopic domain can be studied by observing the change in the density ${}^{n}f$ with increasing time. Using these concepts, the subsequent analysis pertaining to an evolution criterion and a master equation for the types of solid material under investigation may be developed.

3. Evolution Criterion

In the foregoing section the different phase densities have been defined and the problem of whether or not the system evolves to an equilibrium distribution when the boundary conditions holding it in another equilibrium state are suddenly changed may be investigated. Investigations of this type have been made by P. Glansdorff and I. Prigogine [5], F. Schlögl [6] and in a related sense by J. Meixner [7]. For the model of structured materials being discussed in this paper an evolution criterion is established, by what amounts to a coarse graining of time, in the following manner.

The macroscopic domain is considered to have been held for a long time in a stationary thermodynamic state which in this analysis will be known as the "undeformed state". At the time $t = 0$ the thermodynamic boundary conditions are "instantaneously" changed to new time independent values known as the "deformed state". The entire macrodomain then evolves with time to another stationary thermodynamic state determined by these new applied boundary conditions. Corresponding to the initial undeformed state, at time $t = 0^-$, a statistical ensemble of the thermodynamic states of a large number of mesoscopic regions may be defined in Γ-space with a fine grained density ρ_o . On the other hand, observations of the state parameters $\underset{\sim}{\nu}_o^\alpha$ determine the relative statistical weights of the various stars ${}^n\Omega$ in Γ-space resulting in an initial course grained density ${}^n f_o$. The latter, of course, is determined from the occupation numbers $\{{}^\ni n_o\}$ in μ-space. By definition, ρ_o must be constant over each star and equal to these statistical weights ${}^n f_o$ i.e.:

$$\rho_o = {}^n f_o \tag{3.1}$$

At this point in the analysis a suitable measure for discussing the evolution of the system for time $t > 0$ must be introduced. Following F. Schlögl [6] the "information measure" may be defined in the present case either form:

$$i_o = \sum_n {}^n f_o \, \ln {}^n f_o \, M({}^n\Omega) \, , \tag{3.2}$$

$$= \int_\Gamma {}^n f_o \, \ln {}^n f_o \, d\Gamma \, , \tag{3.3}$$

which, for equilibrium only, may be written as:

$$i_o = \int_\Gamma \rho_o \ln \rho_o \, d\Gamma \, . \tag{3.4}$$

The concept of information has been developed from initial work carried out by C.E. Shannon and W. Weaver [8] and A. Rényi [9] with discussions as to its implications being given, for example, by R. Brillouin [10] and M.S. Pinsker [11]. Using this measure the time evolution may be followed with the understanding that the results will be qualitative only.

The fine grained Γ-space density ρ varies in accordance with Liouville's equation:

$$\frac{D\rho}{Dt} = 0, \tag{3.5}$$

where D/Dt is the convective time derivative following the motion of the phase points in Γ-space. This is a completely reversible and phenomenological equation and hence the time evolution of the fine grained density can always be found from this deterministic equation. However, the course grained density follows no such equation and hence at any time t later than $t=0$, ρ_t and ${}^{n}\rho_t$ will be different. The reason for this may be visualized in a manner similar to that given by N.G. van Kampen [12] by noting that, even though the representative points of the state of the mesoscopic regions initially in the same star ${}^{n}\Omega$ always occupy a constant volume in Γ-space equal to $M({}^{n}\Omega)$, the shape of the volume during evolution becomes different from that of the initial volume in such a way that it extends over neighbouring stars ${}^{m}\Omega$. Each of these stars is therefore occupied at time t by filaments of representative points corresponding to a variety of values of the fine grained density ρ. Thus, at any time $t > 0$:

$$\rho \neq {}^{n}f_t , \tag{3.6}$$

and:

$$i_t = \int_\Gamma {}^{n}f_t \, \ln {}^{n}f_t \, d\Gamma = \langle \ln {}^{n}f_t \rangle = \int_\Gamma \rho_t \, \ln {}^{n}f_t \, d\Gamma , \tag{3.7}$$

where $\langle \cdot \rangle$ indicates the statistical mean.

As a first step in establishing the evolution criterion rhe difference between i_0 and i_t for $t>0$, is considered so that:

$$i_0 - i_t = \int_\Gamma \rho_0 \ln \rho_0 \, d\Gamma - \int_\Gamma \rho_t \ln {}^{n}f_t \, d\Gamma . \tag{3.8}$$

However, by using the Liouville equation (3.5) the first term on the right hand side of (3.8) may be replacced by:

$$\int_\Gamma \rho_0 \ln \rho_0 \, d\Gamma = \int_\Gamma \rho_t \ln \rho_t \, d\Gamma , \tag{3.9}$$

which permits the difference in the information measure to be written as:

$$i_0 - i_t = \int_\Gamma \rho_t \ln \frac{\rho_t}{{}^{n}f_t} \, d\Gamma . \tag{3.10}$$

Only in the case where the deformed state coincides with the undeformed state will this difference remain zero. It is further observed, by utilizing relation (2.6), that (3.10) may be expressed by:

$$(3.11) \qquad i_o - i_t = \int_\Gamma \rho_t \ln \frac{\rho_t}{{}^n f_t} - \int_\Gamma \rho_t \, d\Gamma + \int_\Gamma {}^n f_t \, d\Gamma ,$$

$$= \int_\Gamma \rho_t \left(\frac{{}^n f_t}{\rho_t} - 1 - \ln \frac{{}^n f_t}{\rho_t} \right) d\Gamma .$$

Based on the inequality :

$$(3.12) \qquad x \geqslant 1 + \ln x ,$$

it then follows from (3.11) that:

$$(3.13) \qquad i_o - i_t \geqslant 0$$

This inequality not only establishes a first step in deriving an evolution criterion for structured media but, more important, it introduces the concept of "irreversibility" into the thermodynamics of such solids. It is obviously equivalent to the Boltzmann's "Generalized H-Theorem" of classical mechanics.

A second step leading to the evolution criterion concerns the two equilibrium distributions for the course grained density ${}^n f$. In the present formulation there are two equilibrium distributions of the state parameter vector $\underset{\sim}{\nu}^\alpha$ which correspond to the undeformed and deformed state of the macroscopic domain. These distributions are designated by ${}^n f_o$ and ${}^n f_\infty$, respectively. Since it has already been established that the information measure is a suitable parameter for examining these densities and that it always reduces during the evolution of a system, it follows then that the equilibrium distributions should be those that minimize the information subject to the constraints imposed by the boundary conditions. Carrying out the analysis for the deformed state, its equilibrium distribution may be determined by noting that at time $t = \infty$ the expected value of the state vector $\underset{\sim}{\nu}^{\alpha}$ is given by either:

$$\langle \underset{\sim}{\nu}_{\infty} \rangle = \sum_{n} {}^{n}f_{\infty} \, {}^{n}\underset{\sim}{\nu} \, M({}^{n}\Omega) , \tag{3.14}$$

or

$$= \int_{\Gamma} {}^{n}f_{\infty} \, {}^{n}\underset{\sim}{\nu} \; d\Gamma , \tag{3.15}$$

where ${}^{n}\underset{\sim}{\nu}$ denotes the numerical values of the state parameters associated with the star ${}^{n}\Omega$ in Γ-space.

To determine the deformed equilibrium distribution density ${}^{n}f_{\infty}$ it is assumed that to a small variation $\delta\, {}^{n}f_{\infty}$ of ${}^{n}f_{\infty}$ there corresponds a small variation δi_{∞}, given by:

$$\delta i_{\infty} = \int_{\Gamma} \left(\ln {}^{n}f_{\infty} + 1 \right) \delta\, {}^{n}f_{\infty} \, d\Gamma . \tag{3.16}$$

However, this variation is subject to the constraints expressed by the relations (2.7) and (3.15), and hence, by using the method of Lagrangian multipliers these constraints are taken into consideration by:

$$\delta i_{\infty} = \int_{\Gamma} \left(\ell n\, {}^{n}f_{\infty} + 1 \right) \delta\, {}^{n}f_{\infty}\, d\Gamma + \beta_{\infty} \delta \left\{ \int_{\Gamma} {}^{n}f_{\infty}\, d\Gamma - 1 \right\}$$

(3.17)
$$+ \underset{\sim}{\lambda}_{\infty} \cdot \delta \left\{ \int_{\Gamma} {}^{n}f_{\infty}\, {}^{n}\underset{\sim}{\nu}\, d\Gamma - \langle \underset{\sim}{\nu}_{\infty} \rangle \right\} .$$

$$\delta i_{\infty} = \int_{\Gamma} \left(\ell n\, {}^{n}f_{\infty} + 1 + \beta_{\infty} + \underset{\sim}{\lambda}_{\infty} \cdot {}^{n}\underset{\sim}{\nu} \right) \delta\, {}^{n}f_{o}\, d\Gamma .$$

In (3.17) β_{∞} and $\underset{\sim}{\lambda}_{\infty}$ are the undetermined multipliers, $\underset{\sim}{\lambda}_{\infty}$ being an r -dimensional vector. If the expression in (3.17) is to be an extremal value, as it must be in an equilibrium state, the variation δi_{∞} must be zero. This implies that:

(3.18)
$${}^{n}f_{\infty} = \exp\left(-1 - \beta_{\infty} - \underset{\sim}{\lambda}_{\infty} \cdot {}^{n}\underset{\sim}{\nu} \right) .$$

The undetermined multipliers are found by substituting (3.18) into (2.7) and (3.15). Hence, from (2.7):

(3.19)
$$-\varphi_{\infty} = 1 + \beta_{\infty} = \ell n \left\{ \int_{\Gamma} \exp\left(-\underset{\sim}{\lambda}_{\infty} \cdot {}^{n}\underset{\sim}{\nu} \right) d\Gamma \right\} ,$$

where φ_{∞} is a potential function associated with $\underset{\sim}{\nu}^{\alpha}$, whilst from (3.15):

(3.20)
$$\langle \underset{\sim}{\nu}_{\infty} \rangle = \int_{\Gamma} \exp\left(\varphi_{\infty} - \underset{\sim}{\lambda}_{\infty} \cdot {}^{n}\underset{\sim}{\nu}_{\infty} \right) {}^{n}\underset{\sim}{\nu}\, d\Gamma ,$$

$$= \left\{ \int_{\Gamma} \exp\left(-\underset{\sim}{\lambda}_{\infty} \cdot {}^{n}\underset{\sim}{\nu}\right) d\Gamma \right\}^{-1} \int_{\Gamma} \exp\left(-\underset{\sim}{\lambda}_{\infty} \cdot {}^{n}\underset{\sim}{\nu}\right) {}^{n}\underset{\sim}{\nu}\, d\Gamma ,$$

$$= - \frac{\partial \varphi_{\infty}}{\partial \underset{\sim}{\lambda}_{\infty}} . \tag{3.20}$$

Hence, the undetermined multipliers may be determined from (3.19) and (3.20). The deformed equilibrium state has, therefore, the density distribution:

$${}^{n}f_{\infty} = \exp\left(\varphi_{\infty} - \underset{\sim}{\lambda}_{\infty} \cdot {}^{n}\underset{\sim}{\nu}\right) . \tag{3.21}$$

Again making use of (2.7) and (3.15) the information contained in the deformed state is given by:

$$i_{\infty} = \int_{\Gamma} {}^{n}f_{\infty} \; \ln {}^{n} f_{\infty} d\Gamma = \varphi_{\infty} - \underset{\sim}{\lambda} \cdot \langle \underset{\sim}{\nu}_{\infty} \rangle \tag{3.22}$$

In a directly analogous manner the equilibrium density distribution in the undeformed state may be determined as:

$$i_{o} = \varphi_{o} - \underset{\sim}{\lambda}_{o} \cdot \langle \underset{\sim}{\nu}_{o} \rangle \tag{3.23}$$

and hence the "loss of information" between these two equilibrium states is given by:

$$i_{o} - i_{\infty} = \left(\varphi_{o} - \varphi_{\infty}\right) - \left(\underset{\sim}{\lambda}_{o} \cdot \langle \underset{\sim}{\nu}_{o} \rangle - \underset{\sim}{\lambda}_{\infty} \cdot \langle \underset{\sim}{\nu}_{\infty} \rangle\right) \tag{3.24}$$

This change in information completely ignores the evolution path that the process follows in going from the undeformed to the deformed state. It is equivalent to the statement of Clausius concerning the entropy difference between the two equilibrium states.

The third and final step leading to the evolution criterion governing the change in the information measure during the evolution of the structured media from the undeformed to the deformed thermodynamic state is obtained from the Liouville equation (3.5) and relation (3.10) by noting that:

$$0 \geq i_{t+\Delta t} - i_t$$

$$= (i_o - i_t) - (i_o - i_{t+\Delta t})$$

$$= \int_\Gamma \left\{ \rho_t \ln \frac{\rho_t}{{}^n f_t} - \rho_{t+\Delta t} \ln \frac{\rho_{t+\Delta t}}{{}^n f_{t+\Delta t}} \right\} d\Gamma$$

$$= \int_\Gamma \left\{ \rho_{t+\Delta t} \ln {}^n f_{t+\Delta t} - \rho_t \ln {}^n f_t \right\} d\Gamma \tag{3.25}$$

Now, if the thermodynamic states at the times t and $t + \Delta t$ are assumed to be equilibrium states, which again ignores the precise nature of the evolution between these states, then equilibrium distributions similar in form to that given in (3.21) exist at these times. Substituting these distributions into (3.25), and making use of (2.6), (2.7) and relations similar to (3.15), then:

$$0 \geq i_{t+\Delta t} - i_t \tag{3.26}$$

$$= \int_\Gamma \left\{ \rho_{t+\Delta t}\left(\varphi_{t+\Delta t} - \underset{\sim}{\lambda}_{t+\Delta t} \cdot {}^{n}\underset{\sim}{\nu}\right) - \rho_t\left(\varphi_t - \underset{\sim}{\lambda}_t \cdot {}^{n}\underset{\sim}{\nu}\right)\right\} d\Gamma$$

$$= \varphi_{t+\Delta t} - \varphi_t - \left(\underset{\sim}{\lambda}_{t+\Delta t} \cdot \langle \underset{\sim}{\nu}_{t+\Delta t} \rangle - \underset{\sim}{\lambda}_t \cdot \langle \underset{\sim}{\nu}_t \rangle\right) \tag{3.26}$$

Finally, letting $\Delta t \to 0$, after dividing by Δt, (3.26) becomes:

$$\dot{i} = \dot{\varphi} - \overline{\underset{\sim}{\lambda} \cdot \langle \dot{\underset{\sim}{\nu}} \rangle} \leqslant 0 \tag{3.27}$$

This is the evolution criterion for the materials being considered in this paper. It has been derived in a manner which implies a coarse graining of time. If during the time interval Δt it car be assumed that small deviations from thermodynamic equilibrium occur, this amounts to the assumption that the evolution process is "Markovian". This concept will be used in the next section where a master equation is derived.

4. Master Equation

As stated previously, the evolution of the system with time is of primary interest. Since the possibility of integrating the Liouville equation (3.5) governing the time evolution criterion derived in the previous section is only of a qualitative nature, a "master equation" type of formulation is resorted to as a first approximation leading to quantitative results.

From the coarse grained point of view dealt with in previous sections the evolution of a system may be represented in Γ-space by a diffusion process (see ref. [12]). If at time t the state of a mesoregion is represented by a point in the $^{n}\Omega$ star neighbourhood it may be considered, from experimental observation, to have a "conditional probability" $P_t(^{n}\Omega|^{m}\Omega;\Delta t)$ of being in the state $^{m}\Omega$ at time $t+\Delta t$. This conditional probability has the obvious properties:

$$(4.1)\qquad \begin{gathered} P_t(^{n}\Omega|^{m}\Omega;0) = \delta_{nm}\,;\quad P_t(^{n}\Omega|^{m}\Omega;\Delta t) \geqslant 0\,; \\ \sum_m P_t(^{n}\Omega|^{m}\Omega;\Delta t) = 1, \end{gathered}$$

It is now postulated that the stochastic process, as characterized by the thermodynamic parameters $\underset{\sim}{\nu}^{\alpha}$ describing the state of each microelement, is a "Markovian process". This implies that:

$$(4.2)\qquad P_{t+\Delta t}(^{n}\Omega|^{m}\Omega;\Delta t) = \sum_{\ell} P_t(^{n}\Omega|^{\ell}\Omega;\Delta t)P_{\Delta t}(^{\ell}\Omega|^{m}\Omega;\Delta t)$$

Subtracting $P_t(^{n}\Omega|^{m}\Omega;\Delta t)$ from both sides of (4.2), dividing through by Δt and making use of $(4.1)_{iii}$ this condition for a Markovian process to exist may be written, as $\Delta t \longrightarrow 0$, by:

$$\frac{d\,P_t({}^n\Omega|{}^m\Omega)}{dt} = \sum_{\ell} \left\{ P_t({}^n\Omega|{}^\ell\Omega) \lim_{\Delta t \to 0} \frac{P_{\Delta t}({}^\ell\Omega|{}^m\Omega)}{\Delta t} \right.$$

$$\left. - P_t({}^n\Omega|{}^m\Omega) \lim_{\Delta t \to 0} \frac{P_{\Delta t}({}^m\Omega|{}^\ell\Omega)}{\Delta t} \right\} . \qquad (4.3)$$

The terms in (4.3) written in limit form are the "transitional probabilities" defined through:

$$W({}^\ell\Omega, {}^m\Omega) = \lim_{\Delta t \to 0} \frac{P_{\Delta t}({}^\ell\Omega\,|\,{}^m\Omega)}{\Delta t} , \qquad (4.4)$$

and hence (4.3) may alternatively be written as:

$$\frac{d\,P_t({}^n\Omega|{}^m\Omega)}{dt} = \sum_{\ell} \left\{ W({}^\ell\Omega, {}^m\Omega) P_t({}^n\Omega|{}^\ell\Omega) - W({}^m\Omega, {}^\ell\Omega) P_t({}^m\Omega|{}^\ell\Omega) \right\} \qquad (4.5)$$

Finally, upon multiplying both sides of (4.5) by the probability of finding a mesoregion's state in the star neighbourhood of Γ-space ${}^n\Omega$ defined by relation (2.7), then upon summing up over all the n stars in Γ-space, it follows that:

$$\frac{d\,{}^mP_t}{dt} = \sum_{\ell} \left\{ W({}^\ell\Omega, {}^m\Omega)\, {}^\ell P_t - W({}^m\Omega, {}^\ell\Omega)\, {}^m P_t \right\} \qquad (4.6)$$

where:

$$ {}^{\ell}P_t = \sum_{n} P_t\left({}^{n}\Omega \middle| {}^{\ell}\Omega\right) {}^{n}P_t . \tag{4.7}$$

Equation (4.6) is the master equation governing the coarse grained irreversible evolution of the macrodomain. By this method the first approximation to estimating the response of a macrodomain from a rough knowledge of the microscopic details may be obtained. The entire structural nature of the material under investigation is contained in the transitional probabilities $W({}^{\ell}\Omega, {}^{m}\Omega)$ which may conceptually be determined by a technique similar to that discussed by D.R. Axelrad [2].

Writing (4.6) in operator form, i.e.:

$$ \frac{d\underset{\sim}{P}_t}{dt} = \underset{\approx}{Q}\,\underset{\sim}{P}_t , \tag{4.8}$$

its solution may be expressed in the form:

$$ \underset{\sim}{P}_{t+\Delta t} = \exp\left[\underset{\approx}{Q}\,\Delta t\right]\underset{\sim}{P}_t , \tag{4.9}$$

where the transition matrix is assumed time independent during the time interval Δt (again a coarse graining of time). Hence, conceptually at least, by carefully following a small time interval in any experiment and by transforming these observations from the physical space through the μ-space to the Γ-space in order to obtain an estimate of the transition matrix $\underset{\approx}{Q}$ the evolution of the entire system may be predicted.

REFERENCES

[1] D. R. AXELRAD & L. G. JAEGER, "Random Theory of deformation in Heterogeneous Media". Int. Conf. Solid Mechs., Southampton, England, (1969).

[2] D. R. AXELRAD, "Rheology of Structured Media", Archum. Mech. stosow, 23, 1, 131-140, (1971).

[3] J. W. PROVAN, "Deformation of Arbitrarily Oriented Media", Archum. Mech. stosow. , 23, 2, 339-352, (1971).

[4] P. & T. EHRENFEST (1911), "The Conceptual Foundations of the Statistical Approach in Mechanics". Cornell University Press, Ithaca, (1959).

[5] P. GLANSDORFF & I. PRIGOGINE, "On a General Evolution Criterion in Macroscopic Physics". Physica, 30, 351-374, (1964).

[6] F. SCHLOGL, "On the Statistical Foundation of the Thermodynamic Evolution Criterion of Glansdorff and Prigogine". Annals of Physics, 45, 155-163, (1967).

[7] J. MEIXNER, "Processes in Simple Thermodynamic Materials". Arch. Rational Mechs. Anal., 33, 33-53, (1969).

[8] C. E. SHANNON & W. WEAVER, "The Mathematical Theory of Communication". Univ. Of Illinois Press, Urbana, (1949).

[9] A. RENYI, "Wahrscheinlichkeitsrechnung". VEB Deutscher Verlag der Wissenschaften, Berlin, (1966).

[10] L. BRILLOUIN, "Science and Information Theory". Academic Press Inc., 2nd Edition, (1962).

[11] M. S. PINSKER, "Information and Information Stability of Random Variables and Processes". Holden-Day Inc., San Francisco, (1964).

[12] N. G. VAN KAMPEN, "Fundamental Problems in Statistical Mechanics of Irreversible Processes" in: Fundamental Problems in Statistical Mechanics, Ed.: E.G.D. Cohen, North Holland, (1962).

CONTENTS

MIX
Papier aus verantwortungsvollen Quellen
Paper from responsible sources
FSC® C105338

If you have any concerns about our products,
you can contact us on
ProductSafety@springernature.com

In case Publisher is established outside the EU,
the EU authorized representative is:
Springer Nature Customer Service Center GmbH
Europaplatz 3, 69115 Heidelberg, Germany

Printed by Libri Plureos GmbH
in Hamburg, Germany